Ghislain Comlan Akabassi

Knowledge and phenotypic variability of Picralima nitida in Benin

Ghislain Comlan Akabassi

Knowledge and phenotypic variability of Picralima nitida in Benin

Plant Genetic Resources: Endogenous Knowledge and Sustainable Management

ScienciaScripts

Imprint
Any brand names and product names mentioned in this book are subject to trademark, brand or patent protection and are trademarks or registered trademarks of their respective holders. The use of brand names, product names, common names, trade names, product descriptions etc. even without a particular marking in this work is in no way to be construed to mean that such names may be regarded as unrestricted in respect of trademark and brand protection legislation and could thus be used by anyone.

Cover image: www.ingimage.com

This book is a translation from the original published under ISBN 978-620-2-27961-1.

Publisher:
Sciencia Scripts
is a trademark of
Dodo Books Indian Ocean Ltd. and OmniScriptum S.R.L publishing group

120 High Road, East Finchley, London, N2 9ED, United Kingdom
Str. Armeneasca 28/1, office 1, Chisinau MD-2012, Republic of Moldova, Europe
Printed at: see last page
ISBN: 978-620-6-00319-9

Table of contents :

REPUBLIC OF BENIN

UNIVERSITY OF ABOMEY-CALAVI

FACULTY OF SCIENCE AND TECHNOLOGY

DEPARTMENT OF GENETICS AND BIOTECHNOLOGY

Knowledge and phenotypic variability of *Picralima nitida* in Benin

by:

Ghislain C. AKABASSI

DEDICATION

I dedicate this work to my father Michel AKABASSI and my mother Sidicatou FAGBOHOUN as a sign of love and gratitude

ACKNOWLEDGEMENTS

The realization of this work was only possible thanks to the help of many people, both physical and moral, who, from near and far, spared no effort to put at our disposal, precious time, knowledge and advice. May all these people receive here our deepest gratitudes.

We would also like to thank in particular :

> Professor Brice SINSIN, Director of the Laboratory of Applied Ecology (LEA) and Rector of the University of Abomey-Calavi (UAC), for having accepted to carry out this work in his laboratory. Receive, Mr. Professor, our sincere and deep thanks;

> Pr. Dr. Ir. Achille E. ASSOGBADJO, Full Professor of Forestry (CAMES) for having accepted to supervise this work despite his multiple occupations. Your scientific finesse and rigorous work are the sources of the quality of this document;

> Professors Clément AGBANGLA, Corneille AHANHANZO and Hubert ADOUKONOU-SAGBADJA, responsible for the training in Master of Genetics, Biotechnologies and Biological Resources (MGBRB), for having contributed to our training since the second year of Chemistry Biology and Geology at the Faculty of Science and Technology (FAST) up to the Master. The fruits of your trees planted for more than 5 years are now ripe;

> All the professors of the Master of Genetics, Biotechnology and Biological Resources, in particular Professors Jeanne ZOUDJIHEKPON and Micheline AGASSOUNON for your contributions to our training and your multiple advices during the training;

> Dr. Ir. Elie Antoine PADONOU who spared no effort to contribute day and night to this new scientific fire despite his multiple occupations by his advice in the statistical analysis of data;

> Dr. Tiburce Codjo ODJO for the personal interest he has shown in me since my high school days. Thank you for your permanent assistance. May the Eternal fill you with his grace;

> My father Michel AKABASSI and my mother Sidicatou FAGBOHOUN for the seed that you have sown and maintained;

> All the members of the GBEMAVO family, in particular Juliette GBEMAVO for their moral and technical support;

> Mr. Soumaila Soulé ALAMOU chief village of sobè and Mr. Apollinaire OUSSOU-LIO general secretary of the town hall of avrankou and the president of the NGO GRAB Benin for their attachment. Bless you;

> Mr. Mathieu TOVIEHOU direct assistant of the president of the NGO GRAB Benin for the interest granted to this work. The realization of this work depends on your permanent help. Bless you for it;

> All the traditional healers for having put at our disposal, their precious time, knowledge and advice;

> All the comrades of the 5eme promotion of the master of Genetics, Biotechnologies and Biological Resources ;

> Finally, I would like to thank all those who were involved in the realization of this work.

SUMMARY

Africa, a continent rich in biodiversity, is losing its resources due to the lack of knowledge of their virtues by the local populations. Because of this wealth, its inhabitants are dependent on the resources found in these forests. In addition, they have a lot of endogenous knowledge about the medicinal use of these forest products for their primary health care. The purpose of this study is to provide information on the use value, morphological variability, and germination ability of *P. nitida* seeds in southern Benin. To achieve this, 240 informants, randomly selected from 04 socio-cultural groups namely Fon, Goun, Nago, Aïzo were surveyed. The results reveal that the local populations of the 04 sociocultural groups use the roots, barks, leaves and seeds of *P. nitida* for 21 treatments. *P. nitida* is quite important for the local populations of Ifangni, Adjarra, and Toffo. The seeds of the species are the most commonly used organs, followed by the roots, leaves and bark. The Goun socio-cultural group has a better knowledge of the uses of the species than the Nago, Fon, and Aïzo. There is poor horizontal (between communities) and vertical (between age groups) transmission of available endogenous knowledge. Furthermore, the study of the phenotypic variability of the fruits and seeds of *P. nitida* allowed us to identify 03 fruit morphotypes and 05 seed morphotypes of *P. nitida that are* significantly distinct from each other. Class 2, essentially constituted by fruits from Adjarra (93.75%) is longer, wider, thicker and heavier with respective averages of (13.73 ± 0.70 cm); (11.17 ± 0.68 cm); (9.76 ± 0.50 cm); (743.13 ± 128.10 g). Class 3 constituted by fruits from Adjarra (68.42%) is characterized by shorter fruits with respective averages of (10.30 ± 1.00 cm) (length); (7.88 ± 0.85 cm) (width); (6.87 ± 0.62 g) (weight). In addition, classes 2 and 3, which are mostly made up of fruits from Adjarra, have more aborted seeds and a large number of seeds per fruit. Class 2 has thicker seeds and class 3 has wider and heavier seeds. The longest, widest and heaviest fruits contain the longest, widest and heaviest seeds. Germination of the seed morphotypes on simple substrate without any pretreatment was negative. The species thus has a natural regeneration problem in Benin.

Keywords: Endogenous knowledge, use value, morphological variability, germination, *P. nitida*, morphotypes.

ABSTRACT

Africa continent rich in biodiversity loses its resources for lack of the misunderstanding of their virtues by the local populations. Because of this wealth, his inhabitants are dependent on resources which are in these forests. Furthermore, they detain a lot of endogenous knowledge on the medicinal use of these forest products for their care of primary health. The purpose of the present study is to supply information on the use value, the morphological variability and the capacity in the seeding of the seeds of *P. nitida* in South-Benin. To reach there, 240 informants, chosen in a random way in 04 sociocultural groups to know Fon, Goun, Nago, Aïzo was investigated. The results reveal that the local populations of 04 sociocultural groups use roots, barks, leaves, seeds of *P. nitida* for 21 processings. *P. nitida.* is rather important for the local populations of Ifangni, Adjarra, and Toffo. The seeds of the species are the most used organs followed by roots, leaves and bark. The sociocultural group Goun knows better and possesses a better knowledge of the uses of the species with regard to Nago, Fon, and Aïzo. There is a bad horizontal transmission (between communities) and vertical line (between age groups) available endogenous knowledge. Besides, the study of the phenotypic variability of fruits and seeds of *P. nitida* allowed us to identify morphotypes 03 of fruits and 05 morphotypes of seeds of *P. nitida* significantly different from each other. The class 2 established essentially by fruits from Adjarra (93,75 %) is longer, wider, more thick and heavier with respective averages of (13,73 ± 0,70 cms); (11,17 ± 0,68 cms); (9,76 ± 0,50 cms); (743,13 ± 128,10 g). The class 3 established by fruits from Adjarra (68,42 %) is characterized by fruits shorter with the respective averages of (10,30 ± 1,00 cms) (length); (7,88 ± 0,85 cms) (width); (6,87 ± 0,62 g) (weight). Furthermore, the classes 2 and 3 which are established for most part of fruits from Adjarra have more failed seeds and a significant number of seed by fruit. The class 2 presents the more thick seeds and the class 3 of the wider and heavier seeds. The wider and heavier, the longest fruits contain the wider and heavier, the longest seeds. The seeding of the morphotypes of seed on simple substratum without any pretreatment shows itself negative. The species thus has a problem of natural regeneration in Benin

Keywords: endogenous knowledge, use value, morphological variability, seeding, *P. nitida*, morphotypes.

CHAPTER I: INTRODUCTION

1- Context, problem and justification

The whole world is confronted with the destruction of biodiversity, which is intensifying every day. This problem is particularly acute in sub-Saharan Africa because of population growth, natural disasters and, above all, the lack of knowledge of local populations regarding the use values of species. In Benin, the deforestation rate of 70,000 ha/year between 1990 and 2000 was reduced to 50,000 ha/year after 2000 (FAO, 2011). However, much remains to be done because Africa is the richest continent in terms of biodiversity (IPGRI, 1999) and its inhabitants are dependent on the resources found in the forests. In Africa, south of the Sahara, more than 70% of the population depends on natural resources for their survival (Mahapatra, 2000). Furthermore, in 2002, IPGRI research showed that 80% of local African populations depend on medicinal and food plants for their food and primary health care. This destruction combined with overexploitation of resources is gradually leading to the disappearance of medicinal species. Particular attention must be paid to the knowledge and management of natural species to safeguard endangered medicinal plants. The closest observation is that local communities have more information on wild plants than on cultivated plants (IPGRI, 2002). Through these plants, they are able to meet their family needs and treat many diseases through their active principles. In Benin, we have a multitude of plants, some of which are specialized in the treatment of malaria. It would be important that studies be carried out on these plants for their valorization and importance for a sustainable management. Today, many health organizations such as the World Health Organization (WHO) have recognized that phytotherapy can be effective in treating various diseases. Thus, it has recently devoted enormous attention to *Artemisia annua* and other antimalarial plants, including *Picralima nitida* (Bickii et al., 2007). Thus, this study, which aims to contribute to a better knowledge and management of *P. nitida* in Benin, is of great importance.

P. nitida, a tropical plant of the Apocynaceae family, is present from the Ivory Coast to Uganda and southwards to the D.R. of Congo and the Cabinda region (figure

1). It was first described by Stapf in 1894 who gave it the binomial combination *Tabernaemontana nitida*. Two years later it was described again by Pierre who gave it again the combination *Picralima klaineana* in 1896. In 1910 H. Durand moved the species described by Stapf into the genus Picralima of Pierre and formed a new *combination Picralima nitida* (Akoègninou et *al.*, 2006). In ethnopharmacology, throughout its range, the seeds, bark, roots and leaves of the species have the reputation of being a febrifuge and a remedy for malaria and diabetes (Iwu and Klayman, 1992; Adjanohoun, et *al.*, 1996; Aguwa et *al.*, 2001). In traditional medicine, it is also widely used to relieve various pains (François et *al.*, 1996; Adjanohoun, et *al.*, 1996; Aguwa et *al.*, 2001). Biochemical studies (Kapadia et *al.*, 1993; Fakeye et *al.*, 2000; Ramirez & García-Ribio, 2003; Menzies et *al*, 1998) show that the root bark and fruits of *P. nitida* contain akuammigin, akuammicin, picracin and deacetylpicralin, and the leaves contain akuammin, akuammigin, picraphylline and melinonin A (Kapadia et *al*, 1993; Fakeye et *al.*, 2000; Ramirez & García-Ribio, 2003; Menzies et *al.*, 1998). These alkaloids have a strong sympathomimetic and local analgesic activity and their effects are comparable to those of cocaine and more durable than yohimbine (Raymond-Hamet., 1944; Ansa- Asamoah, & Ampofo, 1986; Fakeye, et *al.*, 2004). The germination work done on *P. nitida* is nothing but pretreatment and in vitro culture of its seeds (Gbadamosi, 2012; 2013). In Benin, scientific knowledge on the use value at the level of local communities, morphological diversity and evaluation of the germination ability of *P. nitida* is not documented. Thus, in a perspective of valorization and sustainable management, the evaluation of this knowledge is essential to understand the causes of the lack of knowledge of the ethnobotanical value of the species. Similarly, the analysis of intraspecific diversity (morphological, genetic) and the ability to germinate are essential for the conservation, improvement and sustainable management of *P. nitida*.

2- Research objectives

General objective

The main objective of this study is to contribute to a better knowledge and management of *P. nitida* in Benin.

Specific objectives

1. Assessing the use values of *P. nitida at the* local community level;
2. Evaluate morphological diversity within the species;
3. Evaluate the germination ability of *P. nitida* seeds.

3- Research Hypotheses

From these specific objectives, the following assumptions follow:

1. The lack of knowledge of the ethnobotanical and/or economic value of *P. nitida* is the main cause of the underutilization of the species in southern Benin;
2. There is morphological variability within *P. nitida* in southern Benin;
3. Morphology has an influence on the germination ability of *P. nitida* seeds.

<u>**Legend**</u>: I Range of *P. nitida*

<u>**Figure 1**</u>: Distribution map of *P. nitida* in Africa

<u>**Photo 1**</u>: Complete morphology of *P. nitida*

<u>**Photo 2**</u>: Branch and fruit of *P. nitida*

Source: http://database.prota.org/PROTAhtml/Picralima%20nitida En.htm.

<u>**Photo 3**</u>: Floral morphology of *P. nitida*

Source: http://database.prota.org/PROTAhtml/Picralima%20nitida En.htm.

<u>Photo 4</u>: Pair of *P. nitida* fruits

CHAPTER II: MATERIALS AND METHODS

2.1. Presentation of the study areas

The present study was conducted in southern Benin mainly in three communes (Adjarra, Ifangni, and Toffo) (Figure 2). These three communes are in the Guinean-Congolese zone located between 6°25' and 7°30' N (Adomou et *al.*, 2005). Several reasons underpinned the choice of these communes: the presence of the species, the variability of ethnic groups, and the commercial circuit. The valorization of *P. nitida* requires a good knowledge of these morphogenetic characteristics. Thus, according to the review, it is much more recognized in the Nago/Yorouba, Goun community. Therefore, the choice of Nago/Yorouba and Goun localities will allow us to gather relevant information on the uses of the species. The commune of Toffo was chosen because of the presence of the Fon and Aïzo communities in order to measure their knowledge of the species in relation to other socio-cultural groups.

Table 1: Presentation of the study areas

Study Locations	Location and Area in km²	Climate and rainfall	Relief	Soil	Vegetation	Ethnic groups	populations	Main activities
Ifangni	242	Guinean 1200mm	Very little damage	Ferralitic	Relic of sacred forest and oil palm plantation, the vegetation is varied at the level of the lowlands dominated by hydromorphic species such as raffia palm, banana tree, imperata and fodder	Nago/Holli 64.7%, Yoruba 32.3	71.606 hbts of which 38.174 women and 33,432 men	Agriculture and cross-border trade as an adulterated gasoline
Adjarra	112	Guinean 1200mm	Not very rough	Ferralitic	Sacred forest relic and oil palm plantation	Goun and Fon 83.4 Yoruba 8.2	60.112hbts with 85,08% of young people	Agriculture and cross-border trade as an adulterated gasoline
Toffo	492	Equatorial Guinean 1100mm	Very little damage	Ferralitic	oil palm plantation	Aizo 60% Fon to 30% approx.	74717hbts	Agriculture

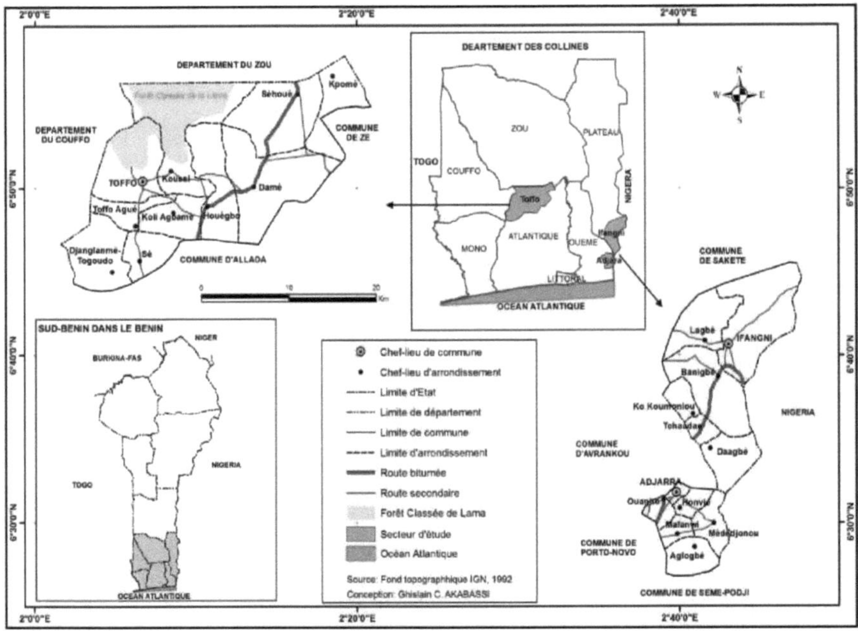

Figure 2: Physical presentation of the villages and towns surveyed

2.2. Ethnobotanical evaluation and use value of *P. nitida* in southern Benin

2.2.1. Sampling

A preliminary study in each zone on the use of the species allowed us to determine the sample size according to the formula of Dagnelie (1998).

$$n = \frac{U^2_{1-\frac{\alpha}{2}} \times p\,(1-P)}{d^2}$$

With n the sample size, p the proportion of respondents who use the species for any purpose (p ≈ 0.1). U1-α/2 =1.96 is the value of the

reduced normal variable fixed for a probability value α = 0,05 and d the margin of error set at 0.08. The individual survey was conducted in each commune and a total of 60 people were interviewed.

2.2.2. Data collection

The survey was conducted in 04 socio-cultural groups: Nago in Ifangni, Goun in Adjarra, Fon and Aïzo in the commune of Toffo. The ethnobotanical study is executed between December 2014 and April 2015 and lasted 15 weeks. Table 2 shows the characteristics of the sample. The interview was conducted in the local language with the help of an interpreter. In each commune, the villages were selected with the help of the secretaries general of the town halls, the arrondissement chiefs and the village chiefs. The survey forms (Appendix 1), designed in French, were used as a basis for the interviews. These forms contain the necessary information on the respondent (name, first name, sex, age, socio-cultural group, and profession), endogenous practices related to the species, the sowing period, the flowering date, cultivation techniques, uses and formulas of use, the various parts used, the frequency of the presence of the species, the frequency of use of the species, etc. The economic value of the species was particularly emphasized during the interview.

Table 2: Characteristics of the sample in each study area

	Socio-cultural groups	Size of the sample	Ages			Sexes	
			18-30 years old	31-60 years old	>61	M	F
Ifangni	Nago/Yoruba	60	20	20	20	30	30
Adjarra	Goun	60	20	20	20	30	30
Toffo	Fon	60	20	20	20	30	30
	Aïzo	60	20	20	20	30	30

2.2. 3. Data processing

The importance of *P. nitida* to the surveyed populations is determined by calculating the frequency of appropriate plant use (FUP) according to Camou-Guerrero et *al.* (2008).

$$FUP = \frac{Rv+Rah+Raf}{Ne} \times 100;$$

With Ne total number of interviewees; Rv, Rah and Raf are respectively number of elderly, adult and young interviewees who use at least one property of *P. nitida*. The property is considered credible if and only if FUP is greater than 50%.

The frequency of specific property use by older men and women (SFUP) and calculated by the following formula:

$$SFUP = \frac{Rv}{Nve} \times 100;$$

With Rv number of elderly men and women who use the property; Nve total number of elderly people. The property is considered to be probably credible if and only if SFUP is greater than 50%.

The overall credibility level of plant properties (CGLP) is assessed by:

$$CGLP = \frac{(Nvc+Nvpc)}{Ntv} \times 100;$$

Where Nvc number of credible properties, Nvpc number of probably credible properties and Ntv total number of properties. The value of CGLP shows the importance of the evaluation of the properties of the species. CGLP < 25%: not very important; 25 < CGLP <50%: weakly important; 50 < CGLP < 75%: somewhat important; 75 < CGLP < 100%: very important.

The most used organ of the species is identified by calculating the index value related to the useful organs (IVO).

With Nvo number of properties related to the organ and Nvt total number of properties

$$IVO = \frac{Nvo}{Nvt} \times 100;$$

identified.

The importance that the different socio-cultural groups give to the properties of the plant is identified by calculating the frequency of use of the properties for each socio-cultural group.

$$FUPE = \frac{Rge}{Ne} \times 100;$$

With Rge number of properties identified in each socio-cultural group and Ne total number of interviewees.

The overall knowledge index (IGKPC) on the species is obtained by :

$$IGKPC = \frac{Vm}{Ne} \times 100;$$

Where Vm is the average total number of properties cited by the elderly, adults and youth. This parameter indicates the overall level of knowledge about the species of ([25]) each socio-cultural group. Excel software was used to code the information collected in the field. A matrix of data related to socio-cultural category, ownership, part used, and user recognition of the species is subjected to principal component analyses (PCA) using minitab16 software to describe the relationship between local knowledge, socio-cultural group, and age class.

2.3. Evaluation of phenotypic variability in fruits and seeds of *P. nitida*

2.3.1. Sampling

In each commune, *P. nitida* plants were identified and fruits were collected at the same time. Table 3 presents the number of plants found in each commune. These plants are sufficiently distant from each other by at least 200 m, which allowed us to avoid collecting fruits from genetically close individuals as recommended by Turnbull (1975). From each plant, fruits were collected from the lower, middle and upper parts as recommended by Tougiani et *al.* The small sample size is explained by the non-availability of the species, the owners' categorical refusal to sell or offer the fruits and the overexploitation of the few available plants.

Table 3: Number of individuals sampled per study area

Zones sampled	Number of villages surveyed	Number of feet found	Number of fruits collected per municipality
Ifangni	15	10	35
Adjarra	15	05	22
Toffo	15	01	01

2.3.2. Data collection

For each fruit, morphometric variables such as length, width, thickness, and weight were measured before sectioning with a knife following the thickness. These descriptors have already been used in other studies by Leakey et *al,* (2000), (2005a, b); Assogbadjo et *al,* (2005); Padonou et *al* (2013; 2014). The total number of seeds, the number of viable seeds and the number of aborted seeds per fruit were counted and labeled. After drying for three days for 4h/day in the sun, for each seed of each fruit, the morphometric characters (length, width, thickness and weight) were also measured. The weight of the seeds was weighed at INRAB using a SARTORIUS electronic scale with a sensitivity of 0 to 220g (precision: 10^{-4}). The measurements of length, width, and thickness were carried out using a decimeter. All the seeds were kept in the laboratory in plastic bags for 2 months before sowing.

Photo 5: Scale used; Source: AKABASSI G.C. (2015)

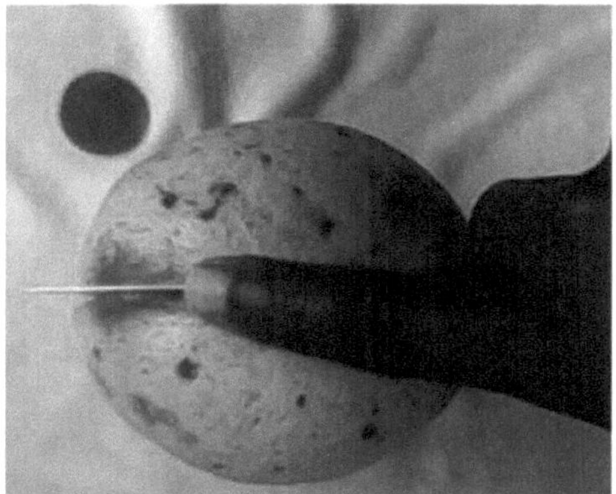

Photo 6: Fruit of *P. nitida* in direct dissection according to thickness

Source: AKABASSI G.C. (2015)

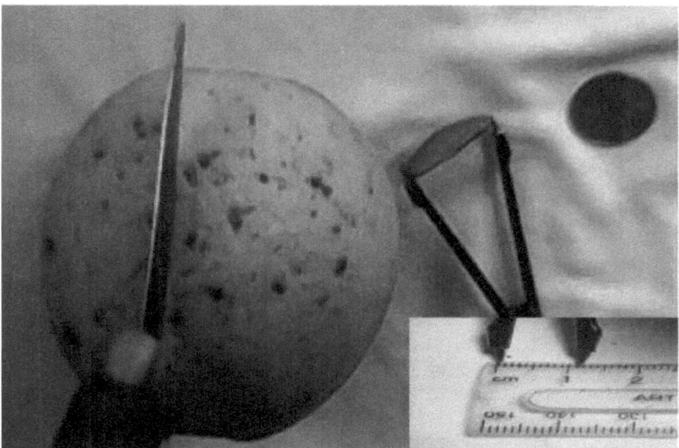

Photo 7: Measurement of morphometric parameters of biological materials

Source: AKABASSI G.C. (2015)

2.3.3. Data processing

After measurements were taken, the collected information on fruits and seeds was coded and subjected to numerical classification to obtain fruit and seed classes. Table 4 shows the coding of the information. A discriminant factor analysis was performed on the identified fruit and seed classes in order to describe them according to their differences (Table 10). The same analysis was also carried out on the morphometric parameters of the seeds according to the communes in order to describe them according to the axes of discrimination (table 13). In addition, intra- and inter-locality variability was assessed using an Analysis of Variance Components (Goodnight, 1978). SAS and minitab14 software were used for the analyses.

Table 4: Coding of variables

Variables	Codes
Length	Long
Width	Larg
Thickness	Epai
Weight	weight
Number of viable seeds	NGV
Number of aborted seeds	NGA
Total number of seeds per fruit	NTG

2.4. Test of the germination ability of *P. nitida* seeds on simple substrate, without any pre-treatment

2.4.1. Experimental device

Germination and growth tests were performed to compare the seed classes identified by the numerical classification. To do this, we adopted an experimental design (Figure 3) in complete randomized blocks with three (03) replications. For each morphotype, 96 seeds were selected at random from the two localities, i.e. 48 per locality. The 96 seeds are distributed in the three (03) repetitions at a rate of 32 per repetition and 16 per locality. All the seeds are previously soaked in water and those which float on the surface of water are considered as seeds having lost their germinative power and are eliminated. Four seeds are sown per pot and a total of 8 pots (4 seed controls with seed coat, 4 with seeds manually stripped of seed coat) are used per class and per replication. A total of 120 pots divided into 3 batches were used for this study. The same substrate was used for both treatments. All pots are filled with sand taken from the FSA application farm. The pot is represented by a polyethylene bag of 18 x 10 x 8 cm^3. The device is watered daily and weekly weedings are performed.

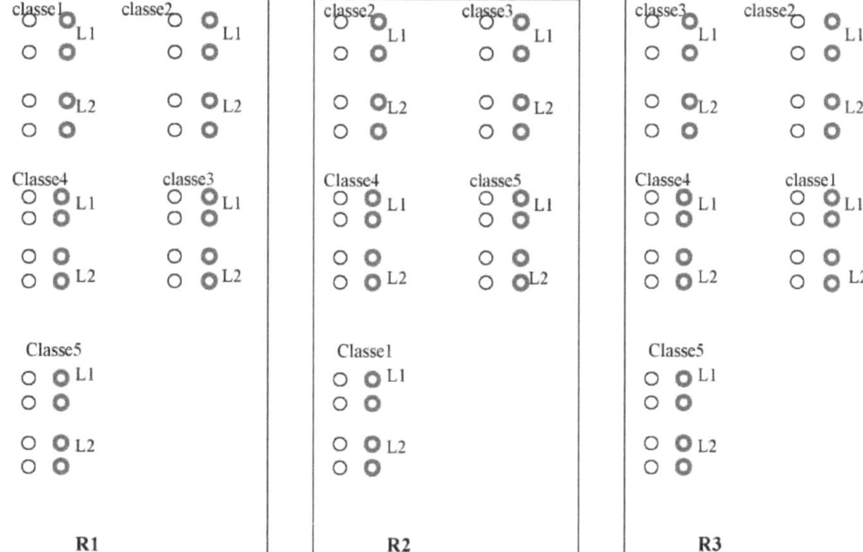

Figure 3: Experimental setup adopted

Legends: L1: locality 1 (Ifangni); L2: locality 2 (Adjarra); ϱ: Seed with seed coat Q: Seed without seed coat; R1: Repeat 1; R2: Repeat 2; R3: Repeat 3

2.4.2. Data collection

Daily the device is checked and the number of germinated seeds is counted over a period of 45 days (duration of seed germination). In the nursery, a seed has germinated when the seedling appears above the substrate (Mapongmetsem, 1994). The time of appearance of the seedling is recorded for each morphotype.

CHAPTER III: RESULTS

3.1. Ethnobotanical evaluation and use value of *P. nitida* in southern Benin

3.1.1. Frequency of *P. nitida* and characteristics of households surveyed in southern Benin

In the three communes sampled, only eight (08) feet of *P. nitida* were found. In the commune of Ifangni with fifteen villages surveyed (Sobè, Monèdjou ègammi, Odofin, Houmbo etc.) only ten plants were found; in the commune of Adjarra with fifteen villages surveyed (Dooké-sedjè, Sèkanmè, Latchè, Gbozounmè etc.) only five plants were found and in the commune of Toffo with fifteen villages only one plant of P. nitida was found.In the commune of Toffo, also with fifteen villages, only one plant of *P. nitida was found* in Agbotagon (Table 5). None of the owners knew the origin of the plants found. According to them, these plants have been there since the time of their ancestors. These owners are essentially traditional healers, traditional practitioners whose knowledge has been handed down to them by their ancestors.

In each municipality, a preliminary study conducted at three levels of scale (ten before the present study, currently and ten after the present study) with a ten-year interval on the perception of local populations relative to the frequency of the species reveals that:

- In Ifangni, out of sixty respondents, 75% think that the species was not very common 10 years ago, 100% think that it is currently very rare, and 95% think that if nothing is done, it will disappear completely by 2025 because it is a plant with high medicinal potential and no protection or popularization measures have yet been taken (Figure 4).

- In Adjarra, statistics show that 85% of respondents believe that the species was uncommon 10 years ago, 90% believe that it is currently very rare, and 100% believe that if nothing is done, it will disappear completely by 2025 because it is a plant with high medicinal potential and no measures have yet been taken to protect or popularize it (Figure 5).

100% believe that if nothing is done, it will disappear completely by 2025 because it is a plant with high medicinal potential and no protection or popularization measures have yet been taken (Figure 5).

- In Toffo, 95% of respondents thought that the species was not very common 10 years ago, 100% thought that it is currently very rare and 100% thought that if nothing is done, it will disappear completely by 2025 because it is a plant with high medicinal potential and no protection or dissemination measures have yet been taken (Figure 6).

Table 5: Number of individuals found per village

Zones sampled	Number of villages surveyed	Number of individuals found	Number of fruits collected per municipality
Ifangni	**15**	**10**	**35**
Adjarra	**15**	**05**	**22**
Toffo	**15**	**01**	**01**

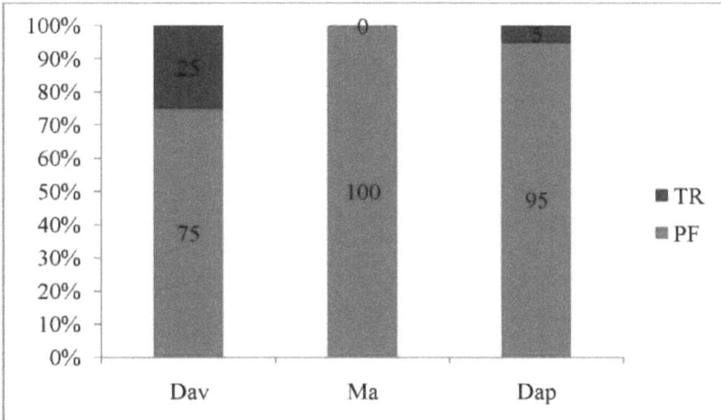

Figure 4: Frequency of *P. nitida* at three levels of scale in the commune of Ifangni Legend: Dav: Ten years ago; Ma: Currently; Dap: Ten years later; TR: Very Rare; PF: Infrequent

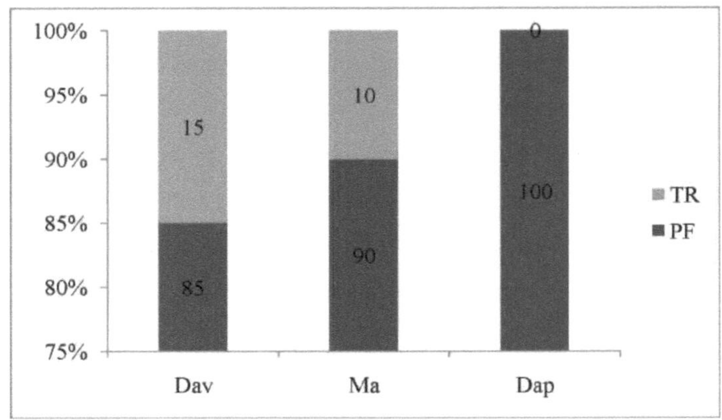

Figure 5: Frequency of *P. nitida* at three levels of scale in the commune of Adjarra Legend: Dav: Ten years ago; Ma: Currently; Dap: Ten years later; TR: Very Rare; PF: Infrequent

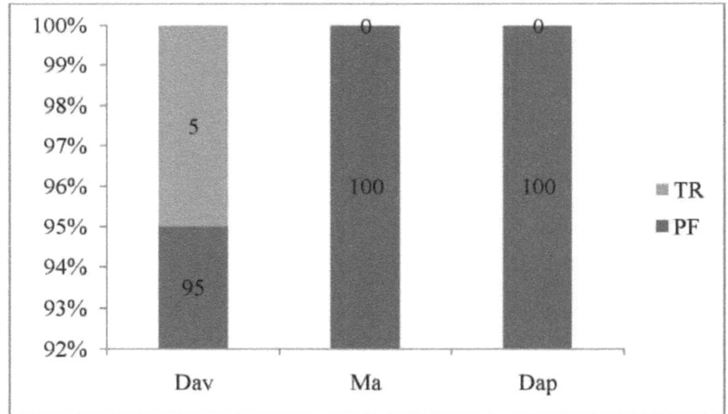

Figure 6: Frequency of *P. nitida* at three levels of scale in the commune of Tofïo

Legend: Dav: Ten years ago; Ma: Currently; Dap: Ten years later; TR: Very Rare; PF: Infrequent

3.1.2. Production and marketing of *P. nitida*

Nago tea sellers in the commune of Ifangni are supplied by two sources, one from Nigeria and the second from Adjarra. Benin has practically no system for producing *P. nitida* fruits. The owners of the few plants found do not market their fruit because the

Seeds are rare and very expensive so they save their fruit for use when needed.

The price of *P. nitida* is almost uniform among all the medicinal plant vendors sampled. The largest fruit (about 960 g) is sold at 200 FCFA each, the medium-sized (about 680 g) at 150 FCFA, and the smallest (about 210 g) at 125 FCFA. After buying the fruits, they break them and let them rot in order to easily collect the seeds. After drying, a seed costs 5 to 10 FCFA. The price of a kilo of *P. nitida* seeds varies between 2000 and 4000 FCFA among all the vendors, but it should be noted that it is more expensive in Toffo.

3.1.3. Ethnobotanical knowledge of *P. nitida* in southern Benin

3.1.3.1.   Use values of *P. nitida* at the level of socio-cultural groups

At the end of our study, it can be noted that knowledge related to the uses of *P. nitida* varies between communities and within each community. The survey reveals that the local populations of the four socio-cultural groups (Fon, Goun, Nago, Aïzo) use the roots, barks, leaves and seeds of *P. nitida* for 21 treatments, including angina, malaria, diabetes, rituals such as burial, marriage, protection, etc...

Table 6 and 7 show the number of uses listed by organ (NUO), the frequency of use (FUP) of the 21 properties of the species among the four socio-cultural groups surveyed, the frequency of specific use of the properties by adults (SFUP), the index value related to the useful organs (IVO) as well as the method of preparation, the dosage and the contraindication of each treatment. It should be noted that 06 treatments are related to the root, 03 to the bark, 04 to the leaf and 17 to the seed. Among the 21 treatments identified, 06 are revealed to be credible (angina, diabetes, malaria, analgesic, stomach ache and ritual practices) 07 are revealed to be probably credible (angina, diabetes, malaria, analgesic, stomach ache, withdrawal and ritual practices).

The overall credibility level of all plant properties (GCLP) is equal to 61.90% which means then that *P.nitida* is quite important for the local populations of Ifangni, Adjarra, and Toffo. The IVO shows that seeds are the most used organs followed by roots, leaves and bark.

Table 8 shows the total average value of uses (Vm) and the overall knowledge index (IGKPC) of *P. nitida for* each socio-cultural group. The table shows that the Goun socio-cultural group has a better knowledge of the uses of the species than the Nago, Fon, and Aïzo.

Table 6: Frequency of use of *P. nitida by* socio-cultural group.

Organs used	the various uses	Number of uses per organ	FUP(%)	SFUP(%)	IVO(%)
Root	-Diabetes		59,58	83,75	
	-Sexual weakness		15,83	30,00	
	-Hemorrhoid		31,25	36,25	28,57
	-Stomach aches	07	54,16	67,50	
	-Toothache		18,33	33,75	
	-Cough		31,25	46,25	
	- Analgesic				
Bark	-Infection of the spleen		15,00	25,00	
	-Sexual weakness	04	15,83	30,00	14,28
	-Infertility		17,91	38 ,75	
	- Analgesic				
Sheet	-Diabetes		59,58	83,75	
	-Malaria	04	75,41	96,25	19,05
	-Cold		02,50	07,50	
	-Measles		06,66	13,75	
Seed	-Angina		100,00	100,00	
	-Diabetes		59,58	83,85	
	-Malaria		75,41	96,25	
	-Analgesic		66,66	100,00	80,95
	-Diarrhea		00,42	01,25	
	-Kwashiorkor	17	02,91	00	
	-Stomach aches		54,16	6,50	
	-Dewormer		23,33	25,00	
	-Measles		06,66	13,75	
	-Infertility		17,91	38,75	
	-Fevers		07,91	05,00	
	-Cough		31,25	46,25	
	-Weaning		30,41	65,00	
	-Cheap		12,08	30,00	
	-Sexual weakness		15,83	30,00	
	-Rituals		66,66	100,00	
	-Against abortion		07,50	25,00	

FUP: Frequency of Use of the Properties of *P. nitida*; **SFUP**: Frequency of Specific Use of the Properties of *P. nitida* by the elderly; **IVO**: Index value related to the useful organs of *P. nitida*.

Table 7: Method of preparation of identified uses

Treatments	Methods of preparation	Dosages	Indications	Other indications
Angina	Suction of the seed	A seed in the morning and a seed in the evening	For people suffering from angina and all other throat pains	-
Analgesic, Stimulant	Decoction of root bark or dried seeds reduced to powder	A glass of bamboo once a day until the pain disappears	Against jaundice, intestinal worms and especially gastric ulcers	-
Against abortion	Decoction of the seed or the boiled leaf	One glass of bamboo morning and evening until the 6ᵉ month of pregnancy	For women in pregnancy and suffering from nausea	Decreased fetal weight
Diabetes	Some seeds in palm wine in a bottle left for 24 hours	A little drink every day	To people suffering from hyperglycemia	No side effects
Diarrhea	Dried seeds powdered in boiled water	A bamboo glass morning noon evening	To those suffering from diarrhea and various pains	forbidden to children less than 10 years old
Hemorrhoid	Decoction of the root	A bamboo glass morning noon evening	For people suffering from internal and external hemorrhoids	No side effects
Malaria	-Crushed leaves + lemon - Boiled leaves with fermented water	A glass of bamboo morning noon evening for 7 days	To people suffering from malaria and especially to children	prohibited for pregnant women
Weaning	Crushed seeds with water on the nipple	-	Weaning children	-
Measles and Fever	Crushed seeds with honey the whole will be passed on the body of the suffering	-	-	-
Cough	Seeds + Salt + Guinea pepper + Honey	For coughs	-	-

| Stomach aches | Boiled root with guinea pepper + small chilli | Stomach disorders and especially gastric ulcer | _ | - |

Table 8: Average total use value (Vm) and index of global knowledge (IGKPC)

Ethnic groups	Vm	IGKPC(%)
Fon	5,58	2,32
Aïzo	4,93	2,05
Goun	8,58	3,58
Nagot/Yoruba	6,51	2,71

on *P. nitida at* the level of each socio-cultural group

3.1.3.2. Link between local knowledge, socio-cultural group and age group

The Principal Component Analysis (PCA) carried out on the different uses of the species, the different parts used, as well as the recognition by the users, showed that the first three axes summarize 61.2% of the initial information, which is sufficient to guarantee an accurate interpretation of the data. The correlation between the variables and the axes (principal components) shows that the following treatments: anti-abortion, weaning, sexual weakness, infertility are well correlated with the first axis. The first axis then describes the use of the medicinal and pharmacological properties of *P. nitida*. Recognition by seed is positively correlated with the second axis while recognition by name and malaria are well negatively correlated with the same axis. The second axis then describes the different forms of recognition of the species by local populations. Root, bark are positively correlated with the third axis while measles, cold, toothache are negatively correlated with the same axis. The third axis then describes the parts used of the plant as well as the medicinal properties related to these various parts.

The projection of the different categories in the plane formed two by two by the three axes (figures n° 7, 8 and 9) made it possible to characterize the different sociocultural groups according to the type of use, the part used and the form of recognition of the species. The elderly Goun and Nago use *P. nitida* more to wean children, to correct infertility in women and sexual weakness in men, while the young people of these different sociocultural groups have none of this knowledge. The young Aïzo, the adults and the elderly Goun recognize the species by the seed, while the elderly Aïzo, Fon, Nago and the adults Aïzo, Fon, Nago recognize the species by the name. Aïzo, Fon, Goun elders and Aïzo, Goun adults make more use of the root and bark to treat illnesses other than measles, colds, and toothache while Aïzo youth, Nago adults and elders have no information on medicinal practices. However, the lack of vertical transmission (adults to youth) of endogenous knowledge among the four socio-cultural groups is a threat to the perpetuation, enhancement, sustainable management, and conservation of the species in these regions.

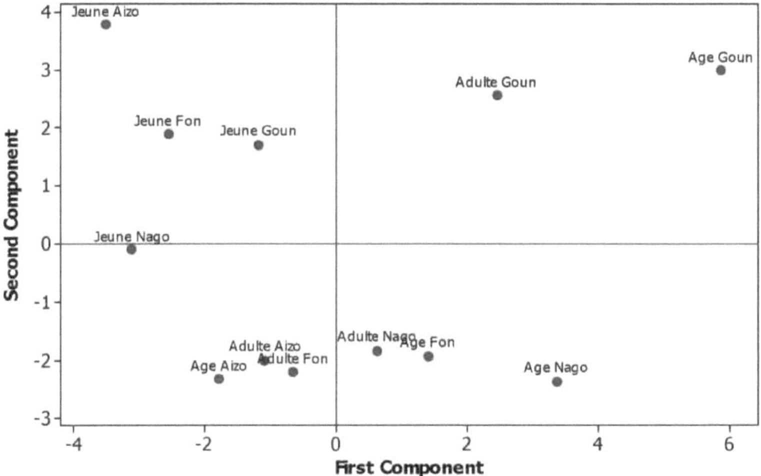

Figure 7: Projection of socio-cultural categories on the first two axes

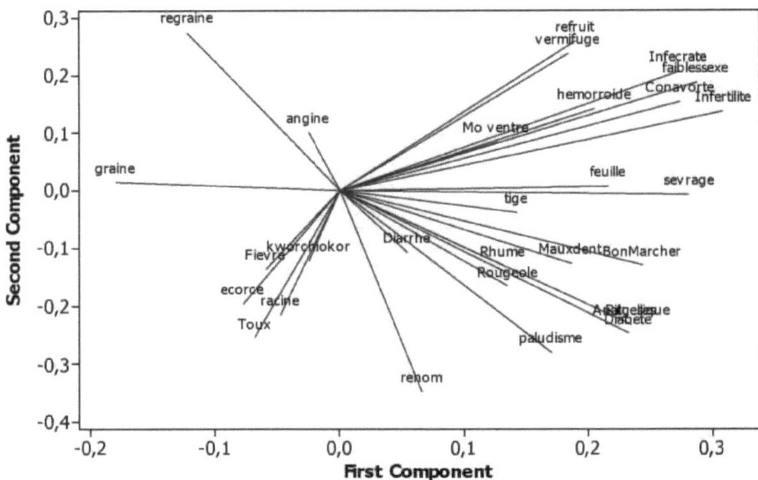

Figure 8: Projection of socio-cultural categories, uses, parties involved and species recognition on the first two axes

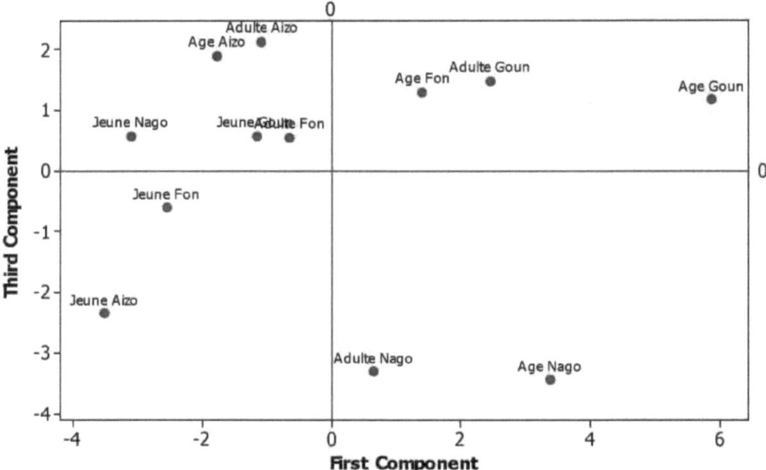

3.2. Phenotypic variability of fruits and seeds of *P. nitida*

3.2.1. Characterization of the fruits of *P. nitida*

The numerical classification carried out in SAS V9.1 on the morphometric variables (length, width, thickness, weight, number of viable seeds, number of aborted seeds, total number of seeds) of the fruits reveals 03 classes of fruits with a coefficient of determination R^2 of 0.65 sufficient for obtaining distinct classes. The dendrogram of figure 10 illustrates the obtained grouping. The results of the discriminant factorial analysis carried out on these different classes show that the first two axes are highly significant and present alone the totality (100%) of the information related to the different classes.

Table 10 shows that the variables length, width, thickness, weight, and NGV are well positively correlated with the first axis with correlations of 0.94; 0.95; 0.97; 0.96; and 0.87 respectively while the variables NGA and NTG are positively represented on the second axis with correlations of 0.80 and 0.71.

Based on Figure 11, the test reveals that, considering axis 1, class 2 fruits are the longest, widest, thickest, heaviest and contain more viable seeds, while class 3 fruits are short, less wide, less thick, lighter and contain few viable seeds. Considering axis 2, the test reveals that fruits of classes 2 and 3 have more aborted seeds and a large number of seeds per fruit while these variables are lower in fruits of class 1.

Table 11 reveals that class 2 gathers the longest, broadest, thickest, heaviest fruits with respective averages of (13.73 ± 0.70 cm); (11.17 ± 0.68 cm); (9.76 ± 0.50 cm); (743.13 ± 128.10 g) while the fruits of class 3 are short, less broad, and less thick with respective averages of (10.30 ± 1.00 cm); (7.88 ± 0.85 cm); (6.87 ± 0.62 cm). Class 2 fruits had the highest average number of viable and total seeds per fruit with average numbers of 53.00 and 73.38 respectively while the lowest average number of viable seeds was obtained by class 3 fruits with an average number of 16.35. The lowest average number of aborted seeds was obtained by class 1 with 3.53. It should be noted that the lowest average total number of seeds is found in class 1 with an average number of 43.16.

The calculation of the percentage of each locality by fruit class represented in Figure 12 shows that all localities are present in all fruit classes. Considering class 1, fruit from Ifangni is predominantly present in the first class with a percentage of 68.42% while it is weakly present in classes 2 and 3 with respective percentages of 6.25% and 4.35%. Fruits from Adjarra are mostly present in classes 2 and 3 with percentages of 93.75% and 95.65% respectively.

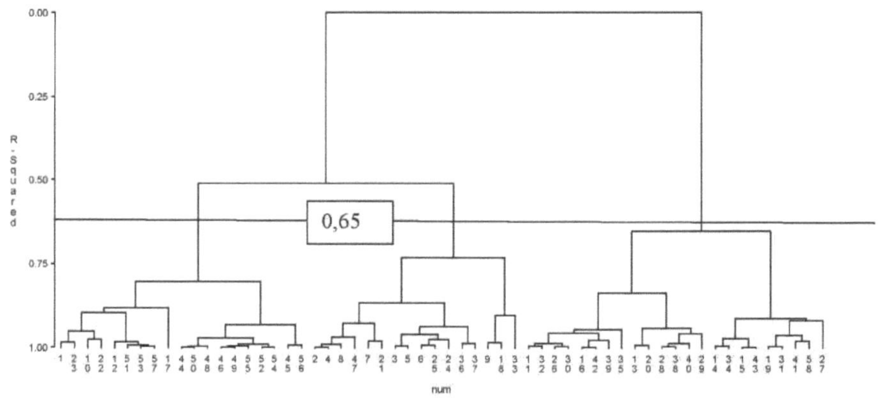

Figure 10: Dendrogram for the fruit grouping of *P. nitida*

Table 9: Mahalanobis distance between fruit classes of *P. nitida*

Classes	1	2	3
1	0ns	-	-
2	8,13***	0ns	-
3	15,82***	36,74***	0ns

ns: not significant, * significant at 0.05; **significant at 0.01; *** significant at 0.001 Prob > Mahalanobis Distance for Squared Distance to CLUSTER

Table 10: Correlation between quantitative morphological descriptors of *P. nitida* fruits and discriminant axes

Morphological descriptors	Axis 1 (0.93) P < 0,0001	Axis 2 (0.07) P < 0,0001
Length (cm)	0,94	-0,07
Width (cm)	0,95	0,03
Epai (cm)	0,97	0,09
Weight(g)	0,96	0,14
NGV	0,87	-0,02
NGA	-0,30	0,80
NTG	0,50	0,71

NGV: Number of viable seeds; NGA: Number of aborted seeds; NTG: Total number of seeds

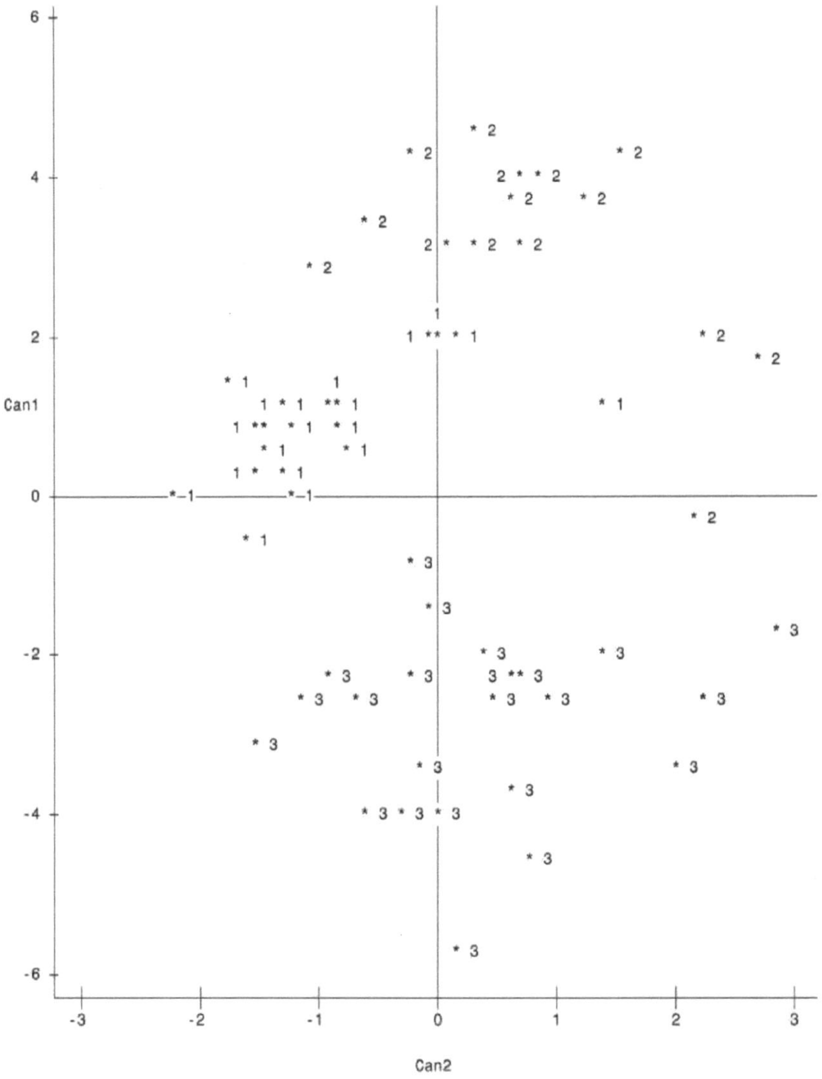

1 : Classe 1 ;   2 : Classe 2 ;   3 :   Classe 3

Figure 11: Representation of the fruit classes of *P. nitida* in the plane

Table 11: Characteristics of the identified fruit classes of *P. nitida*

Variables	Class 1		Class 2		Class 3	
	m	s	m	s	m	s
Length (cm)	12,58	0,52	13,73	0,70	10,30	1,00
Width (cm)	9,85	0,55	11,17	0,68	7,88	0,85
Thickness (cm)	8,52	0,55	9,76	0,50	6,87	0,62
Weight (g)	513,68	111,96	743,13	128,10	244,78	67,48
NGV	39,63	13,39	53,00	14,40	16,35	4,79
NGA	3,53	3,03	20,38	21,61	28,04	18,61
NTG	43,16	14,98	73,38	14,5	44,39	19,81

m: average; s : standard deviation ; NG: number of seed; NT : total number viable; NGA: Number of aborted seeds ; NTG : Total number

NGV: Number of seeds

of seeds

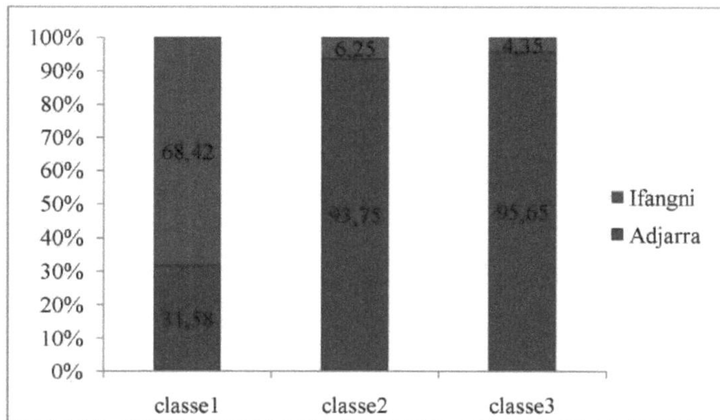

Figure 12: Locations represented by fruit classes of *P. nitida*

3.2.2. Characterization of *P. nitida* seeds

Numerical classification performed with SAS V9.1 software on the morphometric variables (length, width, thickness and weight) of the seeds of these fruit reveals 05 classes of seeds with a coefficient of determination R^2 equal to 0.53 sufficient to obtain distinct classes. The dendrogram in Figure 13 illustrates the obtained grouping. The different seed classes are very significantly distinct from each other ($p<0.0001$) (Table 9). Moreover, the results of the discriminant factor analysis performed on these different classes show that the first two axes are highly significant and present alone 82% of the information related to the different classes.

Table 13 shows that the variables (length and weight) are very well positively correlated with the first axis with respective correlations of 0.73 and 0.95 while only the variable (width) is positively correlated on the second axis with a correlation of 0.82.

Based on Figure 14, the test reveals that, considering the first axis, classes 1 and 3 group the longer, heavier seeds while class 5 groups the shorter, lighter seeds. The classes 2 and 4 contain average values of length and weight. By considering the axis 2 we can conclude that the seeds of the classes 3 and 4 are wider whereas those of the classes 1 and 5 are less wide.

Seeds with the highest mean length value (3.18 ± 0.26 cm) are grouped in class 1. These are followed by the seeds of class 3 which have a higher average value of length ($2,74 \pm 0,34$ cm) whereas the seeds of lower average value of length ($2,18 \pm 0,24$ cm) are grouped in class 5. Seeds with the highest average width values (2.0 ± 0.2 cm) are grouped in class 3 and seconded by those in class 4 with an average width value (1.84 ± 0.14 cm) while seeds with a lower average width value (1.56 ± 0.18 cm) are grouped in class 5. By considering the thickness, it is the seeds of the class 2 which have a greater average value ($0,79 \pm 0,10$ cm) whereas it is the seeds of the class 5 which gathered a lower average value ($0,63 \pm 0,08$ cm). For the variable weight, the seeds of the class 3 are heavier with an average value ($2,02 \pm$

0.28 g) while class 5 has the lightest seeds with a mean value of 0.87 ± 0.30 g (Table 14). In conclusion, the seeds of class 5 have all the weaker variables while those of class 1 are longer, class 2 thicker and class 3 wider and heavier.

Figure 15 shows that both localities are represented in all seed classes. Classes 1, 3, and 5 essentially group seeds from Adjarra (91.43%; 62.25%; 76%). In classes 2 and 4 it is always the origin Adjarra which dominates but this time the difference is not high.

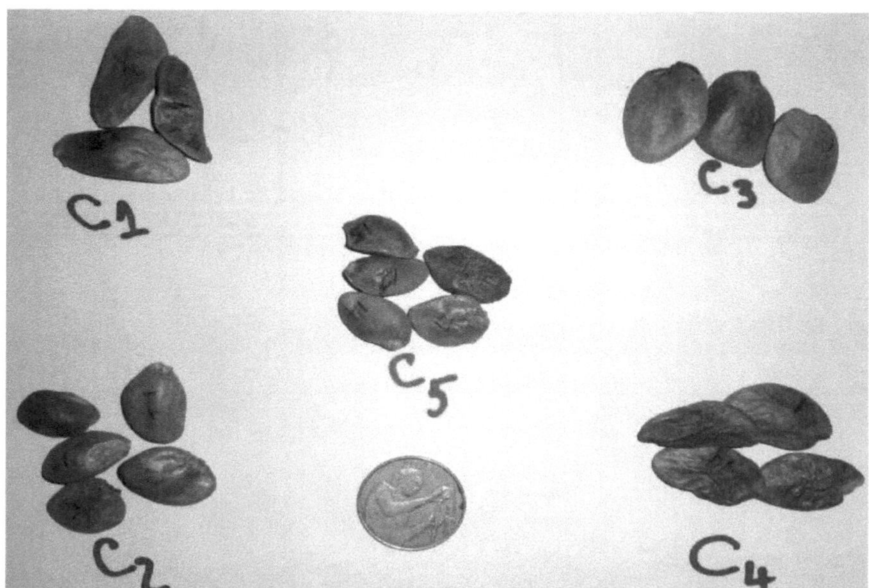

Photo 8: Comparative view of seed morphology of different classes; Source: AKABASSI G.C. (2015)

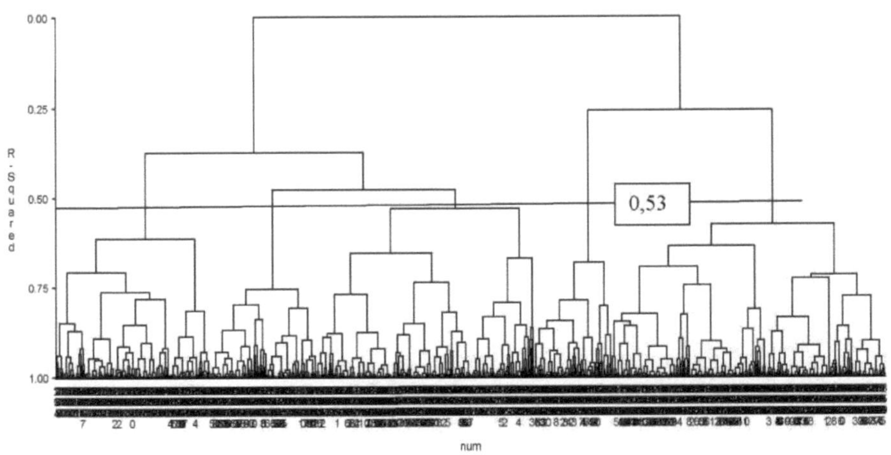

Figure 13: Dendrogram of *P. nitida* seed clustering

Table 12: Mahalanobis distance between seed classes of *P. nitida*

Classes	1	2	3	4	5
1	0 ns				
2	8,55***	0 ns			
3	9 31***	7,10 ***	0 ns		
4	8 77***	6,05 ***	4 29 ***	0 ns	
5	26,39***	12,53***	25,90***	11,56***	0 ns

Prob > Mahalanobis Distance for Squared Distance to CLUSTER ns: not significant, * significant at 0.05; **significant at 0.01; *** significant at 0.001

Table 13: Correlation between quantitative morphological descriptors of *P. nitida* seeds and discriminant axes

Morphological descriptors	Axis 1 (0.54) P < 0,0001	Axis 2 (0.28) P < 0,0001
Length (cm)	0,73	-0,43
Width (cm)	0,51	0,82
Thickness (cm)	0,36	-0,09
Weight(g)	0,95	0,08

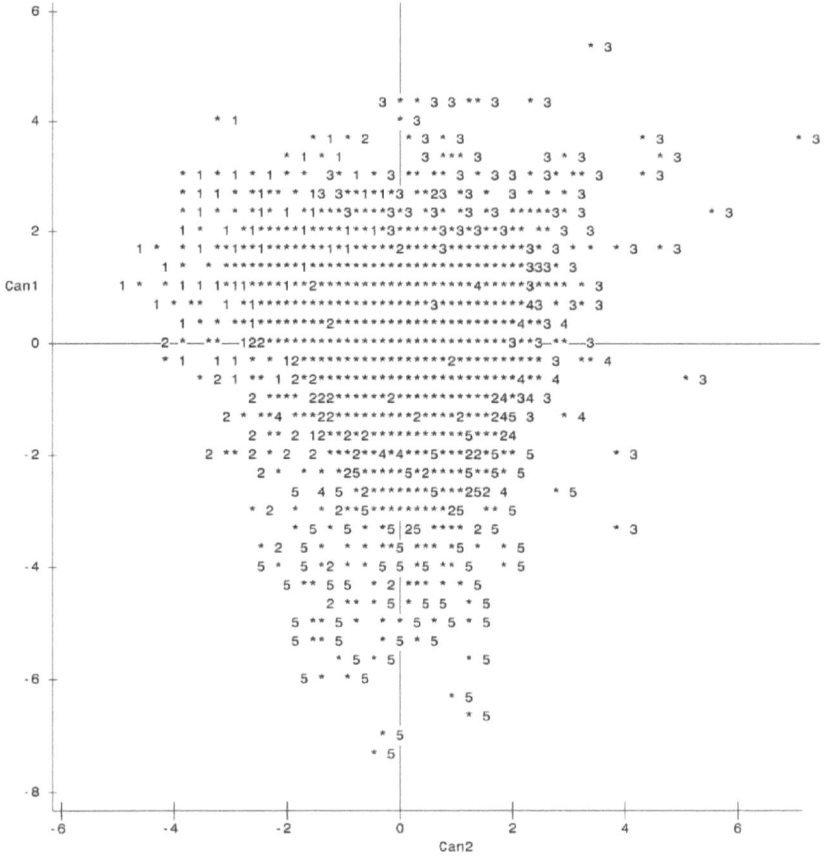

1 : Classe 1 ; 2 : Classe 2 ; 3 : Classe 3 ; 4 : Classe 4 ; 5 : Classe 5

Figure 14: Representation of the seed classes of *P. nititda* in the plane

Table 14: Characteristics of the identified *P. nititda* seed classes

	Class 1		Class 2		Class 3		Class 4		Class 5	
Variables	m	s	m	s	m	s	m	s	m	s
Length (cm)	3,18	0,26	2,39	0,25	2,74	0,34	2,70	0,23	2,18	0,24
Width (cm)	1,60	0,15	1,62	0,15	2,0	0,2	1,84	0,14	1,56	0,18
Thickness (cm)	0,79	0,10	0,86	0,28	0,82	0,9	0,67	0,06	0,63	0,08
Weight (g)	1,9	0,21	1,53	0,28	2,02	0,28	1,71	0,23	0,87	0,30

m: mean; s: standard deviation

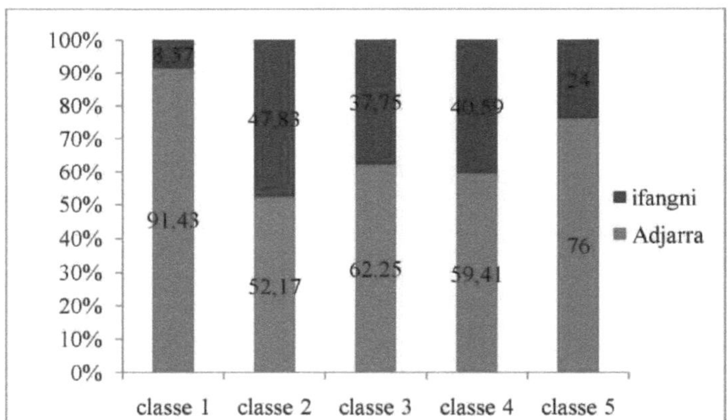

Figure 15: Locations represented by seed class of *P. nititda*

3.2.3. Morphological variation of seeds in relation to fruits of *P. nititda*

The analysis reveals that the first two axes alone contain 100% of the information (Table 15). The different seed classes are very significantly distinct from each other. The variables length and weight are very well positively correlated with the first axis (0.93; 0.75) while only width is positively correlated with the second axis (0.50) (Table 15).

Seeds from G2 class fruits are very long and heavy while those from G3 class are short and light considering the first axis. Based on the second axis, the fruits of classes G1 and G2 contain the widest seeds while the fruits of class G3 are less wide (Figure 16). In conclusion we can say that the longest, widest and heaviest fruits contain the longest, widest and heaviest seeds while the shortest, less wide and lighter fruits contain the shortest, less wide and light seeds.

Table 17 confirms that the longest and heaviest seeds are grouped in class 2 fruits with averages of 2.96 ± 0.37 cm and 1.90 ± 0.36 g while class 3 fruits contain the shortest and lightest seeds with averages of 2.29 ± 0.24 cm and 1.34 ± 0.40 g. The fruits of class 1 group the widest seeds 1.77 ± 0.23 cm while the fruits of class 3 group the least wide seeds 1.67 ± 0.20 cm. Considering the thickness it is necessary to notice that the fruits of the class 3 contain the thickest seeds 0,82 ± 0,15 cm while the fruits of the class 2 gather the least thick seeds.

Despite the significant differences between morphotypes indicated by discriminant analysis, analysis of variance components (Table 18) reveals that variation within localities is as considerable as that between them, for all quantitative traits. In general, 56-100% of the morphological variation is present within localities. However, a significant amount of variation between localities is observed in variables such as number of aborted seeds (42.77) and total number of seeds (43.15).

Table 15: Correlation between seed variables according to fruit class and axes

Morphological descriptors	Axis 1 (0.94) P < 0.0001	Axis 2 (0.6) P < 0.0001
Length (cm)	0,93	-0,32

	0,20	0,50
Width (cm)	0,20	0,50
Thickness (cm)	-0.29	-0,1
Weight(g)	0,75	0,42

Table 16: Mahalanobis distance between fruit classes of *P. nititda*

Classes	1	2	3
1	0ns	-	-
2	1,22***	0ns	-
3	219***	571***	0ns

ns: not significant; *** significant at 0.001 Prob > Mahalanobis Distance for Squared

Table 17: Seed characteristics of the different fruit classes of *P. nititda*

	Group 1		Group2		Group3	
Variables	m	s	m	s	m	s
Length (cm)	2,62	0,31	2,96	0,37	2,29	0,24
Width (cm)	1,77	0,23	1,76	0,26	1,67	0,20
Thickness (cm)	0,78	0,1	0,76	0,11	0,82	0,15
Weight (g)	1,76	0,33	1,90	0,36	1,34	0,40

Mean: average; std: standard deviation

Table 18: Result of variance component estimation procedure on morphological traits of fruits and seeds of *P. nitida* collected in the two localities

Variance component	Long	Larg	Epai	Weight	NGV	NGA	NTG
Inter Location	1,75	2,46	2,60	0,00	0,00	42,77	43,15
Intra Location	98,25	97,54	97,40	100,00	100,00	57,23	56,85

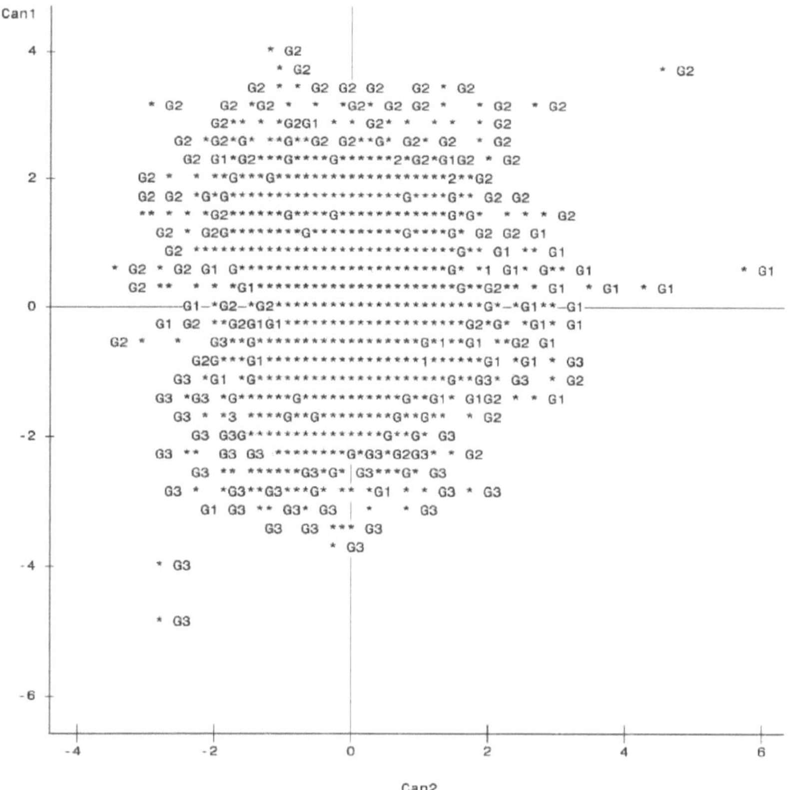

Figure 16: Representation of seeds according to fruit class in the plane

Captions: G1, G2, G3: fruit morphotypes of *P. nititda*

3.3. Test of the germination ability of *P. nitida* on simple substrate, without any pretreatment

In our 45-day germination test, the embryo was unable to open both cotyledons. The manual opening of the cotyledons shows us that germination is not completed before the embryo rots. The same study carried out by changing the substrate (erosion sand) following the same approach gives us the same result. This confirms us that this problem of regeneration of the species is not related to the substrate but can be with the lifespan or the method of conservation of seeds.

CHAPTER IV: DISCUSSIONS

4.1 . Ethnobotanical evaluation and use value of *P. nitida* in southern Benin

To our knowledge and through documentary research, it is for the first time that the use value and the link between local knowledge, socio-cultural group and age class following the variation of socio-cultural groups in Benin are evaluated on *P. nitida.* All the results show that *P. nitida* is a species that can contribute to the economic and health balance. For traditional practitioners, it is used to cure almost all diseases. For herbal tea sellers, it is a source of income (2000-4000FCFA per kilogram). The same observations have been made in other countries through the use of roots, bark, leaves and especially seeds (Iwu and Klayman, 1992; Adjanohoun, et *al.,* 1996; Aguwa et *al.,* 2001; Ansa-Asamoah, & Ampofo, 1986; Keay, 1989). In the localities surveyed, *P. nitida* is a very important plant in the fight against malaria (Iwu and Klayman, 1992).

In addition, the local populations interviewed recognize that the crushing of the parts (seeds, fruits and leaves) of the species gives a very rich lotion to treat various internal and external ailments. This is similar to biochemical studies that show that the fruits of *P. nitida* consist of five alkaloids: akuammidine, akuammine, akuammicine, akuammigine and pseudoakuammigine (Corbett et *al.,* 1996; Menzies et *al.,* 1998; Fakeye et *al.,* 2000; Iwu, & Klayman, 1992; Ramirez & Garcia-Ribio, 2003; Kapadia et *al.,* 1993).

The preliminary study carried out at three levels of scale on the perception of local populations regarding the frequency of the species reveals that the species was infrequent, rare at present and that by 2025 if nothing is done we may see its total disappearance. This is consistent with the work done by Betti (2004) on the same species, which was found to be vulnerable in the Dja biosphere in Cameroon. It is therefore urgent that similar studies be extended to central and northern Benin in order to have a sufficient data base for a conservation program for the species.

In terms of ethnobotanical knowledge, the study revealed that there is a difference between the groups in terms of use value. The Goun have the best knowledge of the uses of the species compared to the Nagos Fons and Aïzo. This result seems normal to us because the popular common name for the species is in Goun (ayokpè); even the Fon and Aïzo communities surveyed in the commune of Toffo only recognize the plant by the local name ayokpè. Moreover, these Goun and Nago populations, being closer to Nigeria, have certainly acquired their knowledge during the fluctuations, given that the plant is more recognized in Nigeria. We believe that it would also be necessary to repeat the same study in the commune of Abomey where the majority of the population is Fon in order to see if the knowledge will eventually be more thorough as some respondents indicated.

This work confirms that the seeds, roots, leaves and bark of *P. nitida* are used by the populations surveyed for 21 treatments, including angina, malaria, diabetes, rituals such as burial, marriage, protection, etc. .... Among these treatments, 06 are revealed to be credible (angina, diabetes, malaria, analgesic, stomach ache and rituals). 07 treatments are revealed to be probably credible (angina, diabetes, malaria, analgesic, stomach ache, weaning and rituals).

The same observations were made by Duwiejuna et *al* (2002); Iwu and Klayman (1992); Keay (1989); Fakeye *et al* (2002). IVO shows that the seeds are used more than the root, leaves and bark respectively. Thus it is without doubt that this is a plant that is really at risk of extinction because the seed is one of the parts where the genetic integrity of plants resides. The overall credibility level of all the properties of the plant (CGLP) is equal to 61.90% means that *P. nitida* has a great medicinal importance for the local populations surveyed. All this information shows that it is urgent to think about the safeguard of this medicinal heritage.

4.2 Evaluation of phenotypic variability in fruits and seeds of *P. nitida*

The present study allowed us to phenotypically characterize the fruits and seeds of *P. nitida* and now to have a database on the morphological variation of *P. nitida* in Benin. The results showed that there is a significant variation (p<0.0001) between the fruits and seeds of *P. nitida* sampled in the two communes of Benin; this allowed us to distinguish 03 fruit morphotypes and 05 seed morphotypes. Other studies already carried out on other species such as *Vitellaria paradoxa, Tamarindus indica,*

Adansonia digitata, Afzelia africana respectively by Maranz and Wiesman (2003); Soloviev et *al.* (2004); and Assogbadjo *et al.* (2011); and Padonou et *al.* (2013; 2014) have shown that morphological characteristics vary well with climate, ecology as well as genetic and abiotic factors

Characterization of the fruit collected in the two communes revealed only three fruit classes, which can be explained by the small sample size. This small sample size can be explained by the rarity of the species, the non-availability of seeds and the categorical refusal of producers to offer or sell seeds. After analysis, the test reveals that class 2, consisting mainly of fruits from Adjarra (93.75%), are longer, wider, thicker and heavier, while class 1, consisting mainly of fruits from Ifangni (68.42%), have intermediate variables. In addition, the analysis of variance components shows that the variation within localities is greater than between localities. This explains that the difference is of genetic origin. However, the environmental origin should not be neglected. These results are similar to those obtained by Assogbadjo *et al.* (2011, 2005a) on the morphological difference of baobab (*Adansonia digitata* L.) fruits in Benin and Padonou et *al.* (2013; 2014) on *Afzelia africana , Jatropha curcas*. Considering the variables NGA (number of aborted seeds) and NTG (total number of seeds), fruits from Adjarra (class 2 and 3) have more seeds per fruit and more aborted seeds while fruits from Ifangni (class 3) contain less seeds per fruit and less aborted seeds. This shows that there is a variation according to origin. These observed variations are probably under the influence of edaphic and especially genetic ecological factors. The high number of aborted seeds in fruits from Adjarra could be explained by genetic drift phenomena such as inbreeding. Since the individuals sampled in Adjarra are located in a forest, the risk of fertilization between genetically similar individuals is high, whereas in Ifangni, the distance between two individuals is high, which limits the risk and consequently low rate of aborted seeds. The few aborted seeds found in Ifangni fruits are the result of self-fertilization. The phenotypic variability observed in the fruits could also be the result of phenotypic plasticity and selection. In addition, sexual recombination, somatic mutations during replication (Kreher et *al.*, 2000; Klekowski, 2003), although rare (Escaravage et *al.*, 1998; Keiper & Mc Conchie,

2000), can increase intra-population genetic diversity.

Analysis of the morphological variation of the seeds according to the fruits reveals that the longest, broadest and heaviest fruits therefore contain the longest, broadest and heaviest seeds while the shortest, less broad and lightest fruits contain the shortest, less broad and lightest seeds. This observation shows a high rate of heritability of morphological traits within the species that could be genetic in nature as a result of meiosis and adaptation to various environmental conditions. It could also be explained by maternal influences. There is therefore a possibility of varietal selection of the species if these genetic variations are confirmed by molecular genetic analysis.

4.3   Germination test of *P. nitida* on simple substrate, without any pretreatment

Through the literature review, no work has yet been done on the germination of the species on simple substrate without any pre-treatment. Moreover, no information has been provided on its natural germination. In Nigeria, the few works done on reproductive biology from seeds are all pretreatments on sterilized substrate and in vitro culture Gbadamosi (2012); (2013). Germination without pretreatment and on simple substrate being less expensive and easy for valorization and conservation, its combination with morphology (morphotype) will allow us to evaluate the influence of morphology on germination. None of the seeds of the 05 classes could express the potential after 45 days. This would have justified the pretreatment work done by Gbadamosi (2012). The complete absence of seedlings under or around the few plants found in the two study areas during the collection phase also justifies that there must be a problem with natural regeneration of the species. This problem may be related to the duration of seed dormancy, the duration of drying, climatic, ecological or edaphic conditions.

Most of the users interviewed during the collection phase do not know the source of the available feet. According to them, these plants date back to the time of their ancestors. In addition, they confirm that *P. nitida* plants are never found in the bush, either in the yard or in the home garden. Scientifically we can relate this fact to the problem of natural regeneration of the species.

Through the bibliography, we found that the seeds of *P. nitida* are highly valued, consumed and disseminated by elephants. In addition, other studies carried out in Côte d'Ivoire by Alexandre (1978) of the Botany Laboratory, O.R.S.T.O.M. Center of Adiopodoumé, on the disseminating role of elephants in the Taï forest, show that the seeds of *P. nitida* found in the dung of elephants have a rather short regeneration period (15 days) and a better growth than those extracted from the fruits and sown in nurseries. This work is similar to that done by Brahmachary (1980) on species such as *Capparis tomentosa* and *Maerua sp* in the Virunga National Park in the former Zaire, where the regeneration of these species could only be found in elephant dung. Thus, the intestinal tract of elephants probably plays an important role in the regeneration of *P. nitida.*

In the light of all the above, it is urgent to continue the work on the natural regeneration of *P. nitida in* order to find a simple and easy method less expensive and accessible to all users for the safeguarding and conservation of *P. nitida* which is nevertheless a species with high medicinal potential.

CONCLUSION AND SUGGESTIONS

Ethnobotanical evaluation through use values, phenotypic characterization of fruits and seeds and germination test on simple substrate without any pretreatment of *P. nitida* seeds were evaluated through this study. It is an important prerequisite for the development of strategies for the valorization and conservation of *P. nitida*. In conclusion, it appears that the species is not well known by the populations; only those who use it know its importance. This is due to a poor distribution of endogenous knowledge between communities and age groups. In addition, the very high medicinal importance of the species within the communities, the poor vertical transmission of information as well as its low presence in the field constitute a favorable basis for its valorization. Moreover, three (03) fruit morphotypes and five (05) seed morphotypes of *P. nitida* are identified through the present study. There is then a great morphological variation of fruits and seeds of *P. nitida* linked to the provenance, potential source of improvement of the species. The non- or low natural germination of the species constitutes a barrier for the study of the influence of morphology on regeneration. This situation can be explained not only by the demographic pressure, the climatic hazards but also by the early loss of the germinative power of its seeds. Faced with these observations, we suggest some urgent avenues of research below:

- To study the vulnerability of *P. nitida in* order to see its status in Benin because it has already proved vulnerable in Cameroon;
- To study the life span and the best method of conservation of *P. nitida* seeds;
- Assessing the genetic diversity of *P. nitida* by molecular and biochemical analysis in relation to its ecology;
- Investigating the correlation between the ecophysiology of *P. nitida* and its adaptive responses;
- Identify the best technique for lifting integumentary dormancy of seeds without pre-treatment through germination tests;
- To develop a program of conservation, improvement and varietal selection of *P. nitida* morphotypes while taking into account the provenance.

REFERENCES

Adjanohoun E.J., Aboubakar N., Dramane K., Ebot M.E., Ekpere J.A., Enow-Orock E.G., Focho D., Gbilé Z.O., Kamanyi A., Kamsu K.J., Keita A., Mbenkum T., Mbi C.N., Mbiele A.L., Mbome I.L., Mubiru N.K., Nancy W.L., Nkongmeneck B., Satabié B., Sofowora A., Tamze V. & Wirmum C.K., 1996. Contribution to ethnobotanical and floristic studies in Cameroon. CSTR/OUA, Cameroon. 641 pp.

Adomou C. A., 2005. Vegetation patterns and environmental gradients in Benin. Implications for biogeography and conservation. PH Thesis; Wageningen University, Wageningen: 133p.

Adoukonou-Sagbadja H., Dansi A., Vodouhe R. & Akpagana K., 2006. Indigenous knowledge and traditional conservation of Fonio millet (*Digitaria exilis* Stapf, *Digitaria iburua* Stapf) in Togo. *Biodiversity and Conservation*, 15, 23792395.

Adoukonou-Sagbadja H., Wagner C., Dansi A., Ahlemeyer J., Daïnou O., Akpagana K., Ordon F. & Friedt W., 2007. Genetic diversity and population differentiation of traditional fonio millet (*Digitaria spp.*) landraces from different agro-ecological zones of from West-Africa. *Theor. Appl. Gen.* 115, 917-931.

Akoègninou A., van der Burg W.J. & van der Maesen L.J.G., 2006. Analytical Flora of Benin. *Backhuys Publishers*, Wageningen.

Alexandre D.-Y., 1978 Laboratoire de Botanique, Centre O.R.S.T.O.M. d'Adiopodoumé: the disseminating role of elephants in the Tai forest.

Anderson Henry T., 1949.The Plant Alkaloids, pp. 759-761, 4th ed, London.

Ansa-Asamoah R. & Ampofo, A.A., 1986. Analgesic effect of crude extracts of *Picralima nitida* seeds. African Journal of Pharmacology 1: 35-38.

Assogbadjo A. E., Sinsin B., Codjia Claude T. J. & Van Damme P., 2005. Ecological diversity and pulp, seed and kernel product of the baobab (*Adansonia digitata* L.) in Benin. Belgian Journal of Botany, 138, 47-56.

Assogbadjo A. E., Kyndt T., Sinsin B., Gheysen G. & Van Damme P., 2006.

Patterns of genetic and morphological diversity in baobab (*Adansonia digitata* L.) population across different climatic zones of Benin (West Africa). Annals of Botany, 97, 819-830.

Assogbadjo A. E., Glèlè Kakaï R., Adjallala Houtoutou F., Azihou F. A., Vodouhê F. G., Kyndt T. & Codjia Claude T. J., 2010. Ethnic differences in use value and use patterns of the threatened multipurpose scrambling shrub (*Caesalpinia bonduc* L.) in Benin; Journal of Medicinal plants Research Vol. 4; November 2010.

Assogbadjo A. E., Glèlè Kakaï R., Edon S., Kyndt T. & Sinsin B., 2011. Natural variation in fruit characteristics, seed germination and seedling growth of *Adansonia digitata* L. in Benin. New Forests 41: 113-125.

Bergonzini Jean Claude, 2004. Climate change, desertification, biological diversity and forests. Paris France, SILVA RIAT 146 pages.

Betti J.L., 2002. Medicinal plants sold in Yaoundé markets, Cameroon. African Study Monographs 23(2): 47-64.

Betti J.L., 2004. An ethnobotanical study of medicinal plants among the Baka pygmies in the Dja biosphere reserve, Cameroon. African Study Monographs 25(1): 1-27.

Bickii J., FeuyaTchouya G.R., Tchouankeu J.C. & Tsamo E., 2007. Antimalarial activity in crude extracts of some Cameroonian medicinal plants. African Journal of Traditional, Complementary and Alternative Medicines, 4(1): 107-111.

Brahmachary R.L., 1980. Germination of seed in the dung balls of the Africa elephant in the Virunga National Park, Zaire. Earth and Life (34): 139 - 142.

Burkill H.M., 1985. The useful plants of West Tropical Africa. 2nd Edition. Volume 1, Families A-D. Royal Botanic Gardens, Kew, Richmond, United Kingdom. 960pp.

Camou-Guerrero A., Reyes-García V., Martínez-Ramos M., Casas A., 2008.

Knowledge and Use Value of Plant Species in a Rarámuri, Community: A Gender Perspective for Conservation. Hum. Ecol. 36: 259-272.

Corbett A.D., Menzies J.R.W., Macdonald A., Paterson S.J. & Duwiejua M., 1996. The opioid activity of akuammine, akuammicine and akuammidine: alkaloids from *Picralima nitida* (fam. Apocynaceae). British Journal of Pharmacology 119: P334Supplement S.

Dagnelie P., 1998. Theoretical and applied statistics. Brussels: De Boeck et Larcier.

Duwiejua M., Obiri D.D., Zeitlin I.J. & Waterman P.G., 1995. Anti-inflammatory activity in extracts from *Picralima nitida* (Fam. Apocynaceae). British Journal of Pharmacology 116: P360 Supplement S.

Duwiejuna M, Woode E, Obiri DD, 2002. Pseudo-akuammigine, an alkaloid from *Picralima nitida* seeds, has anti-inflammatory and analgesic actions in rats. J Ethnopharmacol. 81(1): 73-79.

El-Agamy SZ., 2009. In vitro propagation of some grape root stocks. Acta Hort 839: 125-132.

Escaravage N., Questiau S., Pornon A., Doche B. & Taberlet P., 1998. Clonal diversity in a *Rhododendron ferrugineum* L. (Ericaceae) population inferred from AFLP markers. *Molecular Ecology,* **7,** 975-982.

Ezeamuzie I.C., Ojinnaka M.C., Uzogara E.O. & Oji S.E., 1994. Antiinflammatory, antipyretic and anti-malarial activities of a West African medicinal plant - *Picralima nitida.* African Journal of Medicine and Medical Sciences 23(1): 85-90.

Fakeye T.O., Itiola O.A. & Odelola H.A., 2000. Evaluation of the antimicrobial property of the stem bark of *Picralima nitida* (Apocynaceae). PhytotherapyResearch 14(5): 368-370.

Fandohan B., Assogbadjo A. E., Glèlè Kakaï R., Kyndt T., De Caluwé E., Codjia

J. T. C., Sinsin B., 2010.Women's traditional knowledge , use value and the contribution of tamarind (*Tamarindus indica* L.) to rural household's cash income in Benin. Econ Bot 2010, 64:248-259.

FAO, 2011. State of the world's forests. Food and Agriculture Organization of the United Nations, Rome, 2011.

François G., Ake Assi L., Holenz J. & Bringmann G., 1996. Constituents of *Picralima nitida* display pronounced inhibitory activity esagainsta sexual erythrocytic forms of *Plasmodium falciparum* in vitro. Journal of Ethnopharmacology 54: 113117.

Gbadamosi AE., 2002. Domestication of *Enantia chlorantha* Oliv.- a Medicinal Plant. Ph.D Thesis. University of Ibadan. 192pp.

Gbadamosi AE., Oni O., 2004. *In-vitro* propagation of the endangered medicinal plant *Enantia chlorantha* Oliv. Asset J. 6(2): 27-34.

Gbadamosi AE., 2012. Germination biology of *Picralima nitida* (stapf) under pretreatments Greener Journal of Biological Sciences ISSN: 2276-7762 Impact Factor 2012 (UJRI): 0.7361 ICV 2012: 5.99.

Gbadamosi AE., 2013. In vitro propagation of *Picralima nitida* (stapf) through embryo culture. African Journal of Biotechnology Vol. 12 (22), pp. 3447-3454, 29 May, 2013

Goodnight J.H., 1978. Computing MIVQUE0 estimates of variance components, SAS technical report R-105. SAS Institute Inc, Cary, NC.

Goodson J. A., Anderson Henry T. and Mac Fie J. W. S., 1930. The Biochemical Journal, vol. XXIV, p. 888.

Iwu M.M. & Klayman D.L., 1992. Evaluation of the in vitro antimalarial activity of *Picralima nitida* extracts. Journal of Ethnopharmacology 36(2): 133-135.
Keay RWJ., 1989. Trees of Nigeria: A revised version of Nigerian Trees (1960, 1964). Univ. Press. Pp. 337.

Leakey RB., 1999. Potential for novel food products from agroforestry trees; a review. *Food Chem.* 66 (1): 1-14. Spearson-trustpass.alibaba.com/product/12300525-1137302.

Kapadia, G.J., Angerhofer, C.K. & Ansa-Asamoah, R., 1993. Akuammine: an antimalarial indole monoterpene alkaloid of *Picralima nitida* seeds. Planta Medica 59(6): 565-566.

Keiper F.J. & McConchie R., 2000. An analysis of genetic variation in natural populations of *Sticherus flabellatus* [R. Br. (St John)] using amplified fragment length polymorphism (AFLP) markers. *Molecular Ecology*, 9, 571-581.

Klekowski E.J., 2003. Plant clonality, mutation, diplontic selection and mutational meltdown. *Biol. J. Lin. Soc.* 79, 61-67.

Kreher S.A., Foré S.A. & Collins B.S., 2000. Genetic variation within and among patches of the clonal species, *Vaccinium stamineum* L. *Molecular Ecology*, 9, 12471252.

Leakey RB. 2004. Physiology of vegetative propagation in trees. In: Burley J., Evans E., Younquist J.A. (Eds.), Encyclopaedia of Forest Sciences. Academic Press, London, UK: 1655-1668.

Mapogmetsem P. M., 1994. Phenology and propagation patterns of some agroforestry species in the forest zone. Thesis 3ième Cycle. Univer. Yaoundé I Cameroon, 197p.

Maranz S., Wiesman Z., 2003. Evidence for indigenous selection and distribution of the shea tree, *Vitellaria paradoxa*, and its potential significance to prevailing parkland savanna tree patterns in sub-Saharan Africa north of the equator. J Biogeogr 30:1505-1516

Mathur R. S., Sharma H. K. and Rawat M. M. S., 1984. germination behaviour of provenances of *Acacia nilotica* subsp. indica. Indian Forester, 110, 435-449.

Menzies J.R.W., Paterson S.J., Duwiejua M. & Corbett A.D., 1998. Opioid activity of alkaloids extracted from *Picralima nitida* (fam. Apocynaceae). European JournalofPharmacology350 (1):101-108.

Murashige T, Skoog F., 1962. A revised medium for rapid growth and bio-assays with tobacco tissue culture. Physiol Plantarum. 15:473- 497.

Neuwinger H.D., 1996. African ethnobotany: poisons and drugs. Chapman & Hall, London, UnitedKingdom . 94pp.

Normand D. & Paquis J., 1976. Manual of identification of commercial woods. Tome 2. Afrique guinéo-congolaise. Center Technique Forestier Tropical, Nogent-sur-Marne, France. 335pp.

Obiri D.D., 1997. Studies on anti-inflammatory activity of extracts of seeds of *Picralima nitida*. M. Pharm. Degree thesis, Department of Pharmaceutical Chemistry, Faculty of Pharmacy, Kwame Nkrumah University of Science and Technology, Kumasi, Ghana. 139 pp.

Omino E.A., 1996. A contribution to the leaf anatomy and taxonomy of Apocynaceae in Africa. Wageningen Agricultural University Papers 96-1. Wageningen Agricultural University,Wageningen, Netherlands,178pp.

Padonou A. E., Kassa B., Assogbadjo E. A., Chakeredza S., Babatoundé B. and Glèlè Kakaï R., 2013. Differences in germination capacity and seedling growth between different seed morphotypes of *Afzelia africana* Sm. in Benin (West Africa). Journal of Horticultural Science & Biotechnology (2013) 88 (6) 679-684.

Padonou A. E., Kassa B., Assogbadjo E. A., Fandohan B.,Chakeredza S., Glèlè Kakaï R. and Sinsin B., 2014: Natural variation in fruit characteristics and seed germination of *Jatropha curcas* in Benin (West Africa). Journal of Horticultural Science & Biotechnology (2014) 89 (1) 69-73.

Ramirez A. & Gartfa-Ribio S., 2003. Current progress in the chemistry and pharmacology of akuammiline alkaloids. Current Medicinal Chemistry 10: 18911915.

Raymond-Hamet, 1944. A new sympathicolytic, the akuammidine of Thomas Anderson Henry, thesis doc. med.

SAS Institute Inc. 2003. SAS Online Doc 9.1. SAS Institute Inc, Cary, NC.

Soloviev P., Niang T.D., Gaye A., Totte A., 2004. Variability of physico-chemical characters of fruits of three woody gathering species harvested in Senegal: *Adansonia digitata, Balanites aegyptiaca* and *Tamarindus indica.* Fruits 59:109-119

Tougiani Abasse, John C. Weber, Boubacar Katkore, Moussa Boureima, Mahamane Larwanou & Antoine Kalinganire, 2010. Morphological variation in *Balanites aegyptiaca* fruits and seeds within and among parkland agroforests in eastern Niger.

Turnbull J. W., 1975: Seed collection - sampling consideration and collection tech. In: Report FAO/DANADA Training Course on Forest Seed Collection and handling. Chiang Mai. Thailand. FAO/TF/RAS-11 (DEN). FAO, Rome, Italy. 101 - 122.

http://database.prota.org/PROTAhtml/Picralima%20nitida En.htm

APPENDICES

<u>Appendix 1</u>: Survey Sheet

.... DateCountryDepartmentPrefecture/RegionCity

...... Location/VillageLongitudeLatitude......................

No and first name of the .... informantEthniaSex ....

....................AgeMain activity ....

<u>SPECIFIC OBJECTIVES</u>

What is/are the local name(s) of *P. nitida* .... What does it mean

this name(s) in local .... languageHence

how; by whom; when was it introduced into your locality or household etc.

Are there other plant(s) similar to *P. nitida*.........................

What is/are the difference(s) with *P. nitida*..

Is *P. nitida* considered a plant of great importance to you?

Yes .... No.....................

Does it provide income for local residents or your household?

Nature of income

Economical ....Firewood

.... MedicinalOther...........

Which parts are marketed? And why?

.... LeafStemSeedSeed..............Bark..................................

What diseases does *P. nitida* help to cure? And what are the parts used? Are there

other plants that do the same?

something than *P. nitida?* ....

What are the different uses of *P. nitida*..

And the stakeholdersHow do ....you use them?
....Is there a
.... What is the condition of the
plant?

before; now and what would be its state by 2025 ....

Date: ....Age of plant:.................................... Relief No.: ....

5- Do you know of any place where it exists in large numbers? YES NO . Or is there
a hunter closer to here?

ECOSYSTEM

Number(s) of feet present :

.field :   forest :   house :   other :

Collection site: Sample status: Wild: Spontaneous: No.
of plant sampled :   Number of fruit or flower estimated : Size
        .diameter at 1m from the ground :

            .fruit -size

                -thickness

                -length

                -wide

                -color of the fruit

                -fruit condition: ripe or unripe

                    Dry or wet

            .   Feuille-couleur

                    -situation of the sheet:

                    Young or mature : dry

Number of grains per capsule :

Stress on the collection site

      Abiotic: Biotic:

Are there any other varieties? Yes .... No .... Name them possible ....

What is its flowering date? ....And after which time the fruits are ready for use ....Density of the plant : ....

HighLowAverage

Appendix 2: Correlation Matrix

Variable	PC1	PC2	PC3
Angina	-0.025	0.101	0.266
Mo belly	0.127	0.085	0.016
Dewormer	0.185	0.237	0.133
Cough	-0.069	-0.255	0.118
Hemorrhoid	0.206	0.142	0.243
Malaria	0.170	-0.280	0.185
Conavorte	0.275	0.155	0.049
Weaning	0.281	-0.007	-0.106
Cold	0.132	-0.130	-0.276
Analgesic	0.235	-0.232	0.085
Infecrate	0.274	0.207	0.115
weakness	0.287	0.188	0.079
Infertility	0.309	0.138	0.109

Diabetes	0.233	-0.247	0.034
Mauxdent	0.186	-0.127	-0.374
BonMarcher	0.244	-0.130	-0.213
Measles	0.135	-0.166	-0.374
Fever	-0.059	-0.137	-0.041
kworchiokor	-0.025	-0.122	0.044
Diarrhe	0.055	-0.107	0.105
Rituals	0.235	-0.232	0.085
Seed	-0.180	0.015	-0.117
Sheet	0.216	0.007	-0.193
Stem	0.143	-0.038	-0.096
Root	-0.048	-0.214	0.336
bark	-0.077	-0.195	0.271
renamed	0.066	-0.348	0.101
refruit	0.190	0.260	0.132
see	-0.122	0.275	-0.232

Mo belly =   stomach aches;   Conavorte = against abortion; Infecrate =
infection of the spleen; weakness = sexual weakness; Mauxdent = toothache;
BonMarche = cheap; renom = recognition by name; refruit = recognition by fruit;
regraine = recognition by seed.

<u>Appendix 3</u>: Principal Component Analysis

Eigenanalysis of the Correlation Matrix

Eigenvalue 8.0720 5.7489 3.9271 2.9100 2.1896 1.7574 1.6743 1.3772
Proportion 0.278 0.198 0.135 0.100 0.076 0.061 0.058 0.047 Cumulative 0.278 0.477 0.612
0.712 0.788 0.848 0.906 0.954

VariablePC1PC2PC3PC4
        angine-0,   0250,   1010,266-0,089
Mo belly 0.127 0.085 0.016 -0.104 deworming 0.185 0.237 0.133 -0.101 Cough-0   .069-0
        .2550   .1180   .133
        hemorroide0,   2060,   1420,2430,019
        paludisme0, 170-0,   2800,185-0,078
Convenient 0.275 0.155 0.049 -0.092 **withdrawal 0.281 -0.007 -0.106 0.192** Cold 0.132 -0.130 -
0.276 -0.104 Analgesic0.235-0.2320.0850   .029

        Infecrate0, 2740,   2070,115-0,103

weakness 0.287 0.188 0.079 -0.115
Infertilite0, 3090,   1380,1090,054
Diabete0, 233-0,   2470,034-0,009
Mauxdent 0.186 -0.127 -0.374 -0.039
GoodMarcher 0.244 -0.130 -0.213 0.231
Measles 0.135 -0.166 -0.374 -0.060
Fievre-0, 059-0, 137-0,041-0,409
kworchiokor -0.025 -0.122 0.044 -0.325
Diarrhe0, 055-0,   1070,1050,430
Rituelles0, 235-0,   2320,0850,029
graine-0, 1800,   015-0,117-0,361
leaf0   .216 0.007 -0.193 -0.121
rod 0.143 -0.038 -0.096 -0.390
root-0   .048 -0.2140 .336-0   .059
ecorce-0   .077 -0.1950.271-0.146
renom0   .066 -0.3480 .1010   .018
refruit0   .190 0.260 0.132 -0.064
regraine-0,122 0, 275-0,2320,140

Squared Distance to CLUSTER

	From	CLUSTER123
108.1284815	.81801	
28.	12848036	.73698
315.8180136	.	736980

Because the pooled covariance matrix is singular, the F statistics and p-values are not valid and will not be displayed.

Eigenvalue Difference Proportion			CumulativeRatioF	
Value Num DF Den DF Pr> F	6.1294	0.9269	0.9269	0.08566307
1	6.6545			
20.14	12100 <.0001			
2	0.5251	0.0731	1.00000.65570505	
5.36	551 0.0005			

The CANDISC Procedure

Total Canonical Structure

Variable	Can1	Can2
Long	0.946116	-0.070708
Larg	0.953475	0.038174
Thick	0.969136	0.090498
Weight	0.959533	0.141525
NGViable	0.874125	-0.019643

| NGAvorte | -0.303396 | 0.802365 |
| NTGraine | 0.501668 | 0.708902 |

Appendix 4: SAS results of the seed discriminant analysis

Squared Distance to CLUSTER

5 From CLUSTER	1	2	3	4
1	0	8.55193	9.30926	8.76755
26.38624				
2	8.55193	0	7.10177	6.04801
12.53050				
3	9.30926	7.10177	0	4.29375
25.90325				
4	8.76755	6.04801	4.29375	0
11.55674				
5	26.38624	12.53050	25.90325	11.55674
0				

Prob>Mahalanobis Distance for Squared   Distance to   CLUSTER

From CLUSTER	1	2	3	4
5				
1	1.0000	<.0001	<.0001	<.0001
<. 0001				
2	<.0001	1.0000	<.0001	<.0001
<. 0001				
3	<.0001	<.0001	1.0000	<.0001
<. 0001				
4	<.0001	<.0001	<.0001	1.0000
<. 0001				
5	<.0001	<.0001	<.0001	<.0001
1. 0000				

Squared Distance   to group

From group	G1	G2	G3
G1	0	1.21509	2.19213
G2	1.21509	0	5.71511
G3	2.19213	5.71511	0

Prob>Mahalanobis Distance for Squared Distance to group

From

group   G1   G2   G3

G1   1.0000   <.0001   <.0001

                        G2   <.0001   1.0000   <.0001
                        G3   <.0001   <.0001   1.0000

Eigenvalues of Inv(E)*Hcurrent   row and all that follow are
zero
                        = CanRsq/(1-CanRsq)
                                                            Likelihood
Approximate
                Eigenvalue Difference Proportion CumulativeRatio   F
Value Num DF Den DF Pr> F

        1   0.8989   0.8449   0.9433   0.94330.49961989

137.08   8   2644 <.0001

        2   0.0540   0.0567   1.00000.94872612

23.83   3   1323   <.0001

I want morebooks!

Buy your books fast and straightforward online - at one of world's fastest growing online book stores! Environmentally sound due to Print-on-Demand technologies.

Buy your books online at
www.morebooks.shop

Kaufen Sie Ihre Bücher schnell und unkompliziert online – auf einer der am schnellsten wachsenden Buchhandelsplattformen weltweit! Dank Print-On-Demand umwelt- und ressourcenschonend produzi ert.

Bücher schneller online kaufen
www.morebooks.shop

info@omniscriptum.com
www.omniscriptum.com

Printed by Books on Demand GmbH, Norderstedt / Germany